A GUIDE TO MANAGING TRIBAL BUFFALO HERDS

Written By:
Dr. Heather Hatfield
Dr. Trudy Ecoffey
Mr. Jim Stone

Produced By:
Inter Tribal Buffalo Council
2497 West Chicago Street
Rapid City, SD 57702
www.itbcbuffalo.com

INTERTRIBAL
—BUFFALO—
COUNCIL

ISBN-10: 0989466809
ISBN- 13: 978-0989466806

Acknowledgements

We would first like to thank InterTribal Buffalo Council and the many tribes that we have worked for giving us the opportunity to write this guide.

Dr. Trudy Ecoffey and Dr. Heather Hatfield would also like to thank Dr. Diane Rickerl from South Dakota State University for encouraging the development of this guide. Dr. Rickerl was also helpful in assisting with editing.

A special thank you to Rita Weatherbee, ITBC Administrative Assistant, for assisting with proofreading.

Thank you to United States Department of Agriculture (USDA) and the Small Socially Disadvantage Producer Grant(s) which allowed ITBC to do some of the ground work provided in this management guide.

Thank you to Buffalo Field Campaign for the usage of photos including the cover photo.

Photo taken by Dr. Heather Hatfield, Spirit Lake

Table of Contents

Introduction

Mr. Jim Stone

It is important to note that most Native American people and tribes refer to bison as buffalo. Throughout this document the term buffalo will be used except in direct quotations.

Tribes have played a strong role in the restoration of buffalo in the United States since their near extinction in the 1800s. Tribal restoration efforts have attempted to reintroduce buffalo into the day to day lives of Tribal members, in a manner that is consistent with their historical relationship to the buffalo. While this relationship varies among tribes, the overriding principles are the cultural and spiritual significance of the buffalo and the consumption of buffalo meat. This has been the basis for the creation of all Tribal buffalo herds.

As the restoration efforts have progressed throughout the decades, the manner in which the Tribes have had to manage the buffalo has changed. Early Tribal restoration efforts were impacted by the fencing in of the open range and the battle to eliminate diseases in wildlife and livestock (brucellosis, tuberculosis, etc) in the United States. This effort to eliminate diseases had a domino effect on Tribal buffalo programs. The elimination of disease also eliminated Tribal herds and caused Tribes to have to restart their efforts. Policy changes impacted the manner in which Tribes were able to bring buffalo back to their reservations and how they would manage them once they were there.

The evolution of buffalo management in the United States has led to groups with very different philosophies and management styles. These include: federal government agencies and states attempting to maintain buffalo as a wildlife species; states that classify buffalo as livestock; private producers that have grass fed, all natural approaches to management; and private producers who adopt a livestock management philosophy. Tribes attempt to raise the buffalo as wildlife, while surrounded by land under state jurisdictions that classify them as livestock.

There is a wealth of information, research, books and guides that have been compiled and published that concern themselves with the management of buffalo by the various groups listed above. However, none of these are totally relevant to the Tribal buffalo restoration/management efforts as they do not relate to the special circumstances that surround Tribal buffalo restoration. They do not account for and allow for the cultural relationship and what that entails for the Tribes. This document will attempt to be a management guide for Tribes. This document will be a management guide which incorporates the cultural and biological aspects of Tribal buffalo restoration and the Tribal buffalo programs.

Photo taken by Dr. Heather Hatfield, Nambe Pueblo

"He represents the people and the universe and should always be treated with respect, for was he not here before the two-legged peoples, and is he not generous in that he gives us our homes and our food? The buffalo is wise in many things, and, thus, we should learn from him and should always be as a relative with him" (Brown, 1953, p72).

Background

Mr. Jim Stone

Dr. Heather Hatfield

The buffalo, or bison, have inhabited North America for at least 300,000 years. There is one species now, the American bison (*Bison bison*), which consists of two subspecies, the wood bison and the plains bison. Bison in herds of tens of thousands of animals occupied North America's landscape with wood bison in northern Alaska and the Northwest Territories and the plains bison spanning across North America. Two hundred years ago, the bison population decreased from 30-50 million to roughly 1,000 in 1889. There are currently around 400,000.

When Tribes contemplate the history of buffalo, it takes on a different meaning than strictly the evolutionary history of the species. For Tribes the historical relationship with buffalo is one that is described as being in place since "time immemorial". This phrase reflects the place that buffalo, as well as other animals, played in the creation stories of most Tribes.

The buffalo holds an important place in the histories of the Tribes. In many histories the buffalo was the reason for the existence of man on this landscape. There were covenants made between man and buffalo that the buffalo would provide everything that man needed to survive and in return man would honor the buffalo in his everyday life through use of the buffalo for a means of survival.

The Buffalo was part of us, his flesh and blood being absorbed by us until it became our own flesh and blood. Our clothing, our tipis, everything we needed for life came from the buffalo's body. It was hard to say where the animals ended and the human began.

John Lame Deer

Because of this respected relationship with the buffalo, every part of the buffalo was used for life among tribal people. The animal was not just used for food purposes, nothing was wasted.

> *Rarely has the dependence on a single animal so dominated and shaped human lives. From this magnificent creature, standing up to six feet tall at the shoulder and weighing as much as one ton, the Indians obtained practically every necessity of life. In addition to providing huge quantities of meat and the dung for fuel to cook it, the buffalo yielded hides for raiment and shelter – and brain paste for tanning them – tallow to burn for light, horns for spoons and other utensils, bones for tools, bladders for containers, hair for ropes, and tendons for bowstrings. The Indians even found use for the hoofs, turning them into ceremonial rattles to beseech the sacred powers to ensure successful buffalo hunting.* (The Editors of Time-Life Books, 1993, p7).

In the mid 1800s, the goal of the federal government was to eliminate the buffalo as a way to subdue the Tribes. They came very close to accomplishing both of their objectives. By the late 1800s the herds that once numbered in the millions now numbered in the hundreds. The tribes for the most part, were relegated to reservations where they survived from rations provided by the federal government. With the decline in the buffalo population the lives of many tribal people were severely affected physically, emotionally, and spiritually.

When the buffalo went away the hearts of my people fell to the ground, and they could not lift them again.

Plenty Coups

Photo taken by Buffalo Field Campaign, Yellowstone National Park

Tribal Legislation

Dr. Heather Hatfield

Some Tribal people, seeing what was occurring, attempted to save the buffalo and to expand the numbers through restoration efforts on the reservations. Some of the very first buffalo conservationists were Tribal members. In the late 1800s, tribal families like the Walking Coyote, Pablo and Dupree were first to start looking out for the future of buffalo and to protect small numbers of animals. These first buffalo later became the founding stock for several national parks and refuges including Yellowstone National Park and The National Bison Range (Table 1).

The following Table is taken from Halbert entitled "History of establishment for 11 US federal bison populations" (Halbert, p. 2, 2007). The table outlines the history of federal herds.

Table 1: History of establishment for 11 U.S. Federal Bison Populations

Herd	Location	Year	Founding Stock (number, source)
Badlands National Park	South Dakota	1963	3, Ft. Niobrara National Wildlife Reserve; 50, Theodore Roosevelt National Park
		1983	20, Colorado National Monument
Ft. Niobrara National Wildlife Reserve	Nebraska	1913	6, private ranch, Nebraska; 2, Yellowstone National Park
		1935	4, Custer State Park, South Dakota
		1937	4, Custer State Park, South Dakota
		1952	5, National Bison Range

Grand Teton National Park	Wyoming	1948	20, Yellowstone National Park
		1964	12, Theodore Roosevelt National Park
National Bison Range	Montana	1908	1, Goodnight herd; 3, Corbin (McKay-Alloway); 34, Conrad (Pablo-Allard)
		1939	2, 7-Up Ranch
		1952	4, Ft. Niobrara National Wildlife Reserve
		1953	2, Yellowstone National Park
		1984	4, Maxwell State Game Refuge, Kansas (Jones)
Neal Smith National Wildlife Reserve	Iowa	1996	8, Ft. Niobrara National Wildlife Reserve; 8, Wichita Mountains National Wildlife Reserve
		1997	6, Ft. Niobrara National Wildlife Reserve; 8, National Bison Range
		1998	3, Ft. Niobrara National Wildlife Reserve
Sully's Hill National Game Reserve	North Dakota	1919	6, Portland City Park, Oregon
		1932	1, Wind Cave National Park
		1941-1979	7, Ft. Niobrara National Wildlife Reserve
		1987	3, National Bison Range
		1994-1997	2, Theodore Roosevelt National Park
Theodore Roosevelt National Park	North Dakota	1956 1962	29, Ft. Niobrara National Wildlife Reserve

Wichita Mountains National Wildlife Reserve	Oklahoma	1907	15, New York Zoological Park
		1940	2, Ft. Niobrara National Wildlife Reserve
Wind Cave National Park	South Dakota	1913	14, New York Zoological Park
		1916	6, Yellowstone National Park
Yellowstone National Park	Wyoming, Idaho, Montana	1902	Approximately 30 indigenous; 18, Pablo-Allard herd; 3, Goodnight herd

Although the federal government was building herds, Tribal restoration efforts were derailed by the changing land management policies on the reservations brought on by the Dawes Act. The Dawes Allotment Act came into effect in the year 1887. This act ordered the division of tribal lands and the allotment of them to individual tribal members.

> *The Allotment Act authorized the President to allot portions of reservation land to individual Indians. Allotments of 160 acres were to be made to each head of a family and 80 acres to others, with double those amounts to be allotted if the land was suitable for grazing . . . Finally, and perhaps most notably, the Act authorized the Secretary of Interior to negotiate with the tribes for disposition of all 'excess' lands remaining*

after allotments, for the purpose of non-Indian settlement. (Canby, 1981, p. 20).

The Dawes Act was amended in 1891 and again in 1906 by the Burke Act. The Burke Act amended the sections of the Dawes Act dealing with US Citizenship (Section 6) and the mechanism for issuing allotments. The Secretary of Interior could force the Indian Allottee to accept title for land. US Citizenship was now granted unconditionally upon receipt of land allotment (no longer needed to move off the reservation to receive citizenship). Land allotted to Indians was taken out of Trust and subject to taxation.

Both of these Acts had a devastating impact on the Tribal landscape in a manner and impacted early efforts by Tribal members to restore buffalo on the Reservations. The Dawes Act eliminated the communal holding of property which led to the fencing in or the open range. This then limited the amount of buffalo the individual Tribal member could support. Both Acts led to the loss of Tribal lands, because lands surplus to the Tribal needs were made available for sale to non-Tribal people. And, the Tribal members were able to sell the land they were allotted in fee patent. Land granted to Tribal members in fee patent was also taxable and many Tribal members lost land due to inability to pay taxes. As a result, many individual Tribal members who held buffalo, had to sell the animals they had returned to the Tribal lands.

Lulka writes that, "*Tribal herds developed along a different timeline. The first tribal herd was established on the Crow Indian Reservation in 1935. Eighty three bison were donated to the tribe from Yellowstone National Park. In the same year, the Pine Ridge*

Reservation received nine bison from the same source" (Lulka, 2006, p. 75). However, management practices were changed and public herds did not sell or donate again until the 1970s. Lulka further adds that,

> *The overwhelming majority of bison donated from Badlands National Park went to the Oglala Sioux and the Cheyenne River Sioux. Many of the bison donated from the National Bison Range were given to the Confederated Salish and Kootenai in Montana. In the late 1980s, Wind Cave National Park donated bison to the Lower Brule Sioux and the Oglala Sioux in South Dakota. Theodore Roosevelt National Park initially had a wider scope of dispersal, donating to the tribes such as the Sisseton-Wahpeton (SD), Round Valley (CA), Blackfeet (MT), Fort Berthold (ND), and Jicarilla (AZ). Nonetheless, prior to the formation of the ITBC (Intertribal Bison Cooperative) in 1991, only about 16 tribal herds with 2,800 animals existed.* (Lulka, 2006, p. 75).

In the end, Canby argues that,

> *The primary effect of the Allotment Act was a precipitous decline in the total amount of Indian-held land, from 138 million acres in 1887 to 48 million in 1934. Of the 48 million acres that remained, some 20 million were desert or semi-desert. Much of the land was lost by sale as tribal surplus; the remainder passed out of the hands of allottees. Allottees who received*

patents after 25 years found themselves subject to state property taxation, and many forced sales resulted from non-payment . . . Passage of many of the fee allotments out of Indian hands left large 'checkerboard' areas of alternate white and Indian ownership, making sizable farming or grazing projects impractical. (Canby, 1981, p. 21-22).

In 1992, the Tribes formed the InterTribal Bison Cooperative (ITBC) to assist in the return of buffalo to Tribal lands. In 2010, the organization changed their name to the InterTribal Buffalo Council and became a federally chartered Indian Organization under Section 17 of the Indian Reorganization Act.

Photo taken by Buffalo Field Campaign, Yellowstone National Park

Tribal Land Issues

Dr. Heather Hatfield
Mr. Jim Stone

The Indian Reorganization Act (IRA) of June 18, 1934 (signed by President Franklin Delano Roosevelt (FDR)) the Indian New Deal, was U.S. federal legislation that secured certain rights to Native Americans, including Alaska Natives. These included actions that contributed to the reversal of the Dawes Act's privatization of communal holdings of American Indian tribes and a return to local self-government on a Tribal basis. The Act also restored to Native Americans the management of their assets (being mainly land) and included provisions intended to create a sound economic foundation for the inhabitants of Indian reservations. Under the Act, Tribes reorganized their governments by adopting constitutions that modeled the Tribal governments after the US governmental structure. Some Tribes, however, did not accept the provisions of the IRA and drafted their own constitutions.

The land management issues faced by Tribes were heavily influenced by the above listed Acts. The Acts have resulted in differing forms of government, different ownership patterns of Tribal land, and varying influences on Tribal lifestyle. An impact for the Tribal buffalo programs has been that there is no consistency or pattern for locations of the Tribal buffalo program within Tribal government. In some Tribes it is a stand alone program, or it is a part of the wildlife program, and still others include buffalo programs as part of the livestock program or the natural resources program. The manner in which Tribes restore buffalo is dependent on the Tribal land status. Tribes can have checker-boarded lands (adjoining lands that are Tribal and non-Tribal) which impacts who has jurisdiction; or they may have the opposite, where there are Tribes that have continuous land bases. The Tribes with checker-boarded land have to be more diligent in keeping their animals in the allocated range. Buffalo that go outside the Tribal lands are subject to state jurisdiction, which in most cases gives the adjoining land owner the ability to kill the animals as trespassing livestock. The level to which Tribal members have embraced the practices of ranching is very far reaching in Indian Country, and has led to competition between cattle producers and the Tribal buffalo programs. This makes it difficult for the buffalo programs to get land and expand the Tribal herds. When they are able to get land, they have to pay the going lease rates because that is the only source of revenue for the Tribal governments in most cases.

The restoration of buffalo to the Tribal landscapes faces many issues. There are jurisdictional land issues and there are issues in states that classify buffalo as livestock and as such must follow the states livestock regulations when they are on state governed lands. The Tribes for the most part consider the buffalo as wildlife. Tribes have to be very diligent about keeping their buffalo on Tribal lands and this has led to the Tribal buffalo herds having to be "very highly managed" wildlife.

Tribe's are restoring buffalo to restore traditional relationships and usage of the animals, including consumption. As needs have expanded, Tribes need to increase buffalo production. The number of buffalo that a Tribe is able to raise on their available land resources is sometimes not enough to provide meat for all Tribal members. Lulka states,

While this may be a valid goal, tribes are hindered by the fact that they do not have enough bison to supply such a program by themselves. The exact numbers vary, but tribes own only about 10,000 bison. An increase in the use of these animals could potentially deplete some herds. To change this situation, tribes would either have to increase the intensity of production or gain access to more grazing land. (Lulka, 2006, p.79).

Because of the loss of communal land and the fact that Tribal lease income is usually the only revenue generated by Tribes, access to land for the buffalo programs is limited or has to be paid for by the program. Tribal buffalo programs also have costs for the processing of animals for use by tribal members; and for required infrastructure, equipment, maintenance and staffing required maintaining the herd and access animals as needed.

The maintenance of the Tribes cultural link to the buffalo adds a dynamic to the management philosophies. The tribe must maintain the integrity of the buffalo in a manner that is close to the historic relationship. Because of the diversity of Tribes that have a historical relationship with the buffalo, it is hard to develop standards for measuring the level of cultural usage in Tribal programs.

These are just some of the many management issues that have evolved as Tribes attempt to re-establish their relationship with the buffalo. These also show the differences that exist in the Tribal environment of buffalo restoration and management and the other scenarios that exist.

Current Status of Buffalo in Indian Country

Mr. Jim Stone

ITBC has 58 Member Tribes in 19 states with 50 herds of buffalo numbering 15,000 animals on 1,000,000 acres of Tribal lands. It is estimated that there about 5-10 Tribes with herds that aren't a member of ITBC and between 15-20 individual Tribal members with buffalo herds. The Tribes of ITBC harvest approximately 3,000 animals annually for internal usage and/or sale.

By contrast there were approximately 200,000 buffalo in commercial herds in 2007 and 60,000 buffalo were processed for commercial purposes. The US Fish and Wildlife Service manage 1,100 buffalo on six refuges and the National Park Service manages 4,700 buffalo on 5 National Parks.

Photo taken by Buffalo Field Campaign, Yellowstone National Park

Buffalo Management

Mr. Jim Stone

The main goal of the Tribal Buffalo programs is captured in the Mission Statement of ITBC: *"To restore bison to Indian Nations in a manner that is compatible with their spiritual and cultural beliefs and practices."*

The goal of this handbook is to offer an overview of the history and need for Tribal buffalo herds and to provide management guides. The management guide is based on what the Tribe's expectations are for their herds and will help them meet those needs using the current tools of management.

There are many definitions of sustainability. For Tribal buffalo management, this document defines sustainability as meeting the needs of the present buffalo population without threatening or creating a negative impact on the future needs of people, the environment, or other wildlife populations.

For purposes of the Tribal buffalo programs we will focus on the creation of sustainable programs that consists of the following components:

1. Cultural Sustainability
2. Ecological Sustainability
3. Economic Sustainability

1. **Cultural Sustainability**
Dr. Heather Hatfield

The significance of culture is very important in making sustainable buffalo programs. Even if some tribes do not have buffalo the existence value is still high among people. "*For many people, the value of wildlife goes beyond the opportunity to shoot or photograph, or even see, a particular species. They argue that existence value, based on simply knowing that a species exists, is enough reason to protect and preserve it*" (Cunningham & Cunningham, 2008, p235).

For many tribes, the buffalo were more than an animal, "*but is a spiritual presence connected to the ultimate realities of the world. Bringing the bison home, then, is for Indians above all a cultural goal, and educating young people about bison is a way of educating them about spiritual issues. Bison were, of course, a source of food, robes, tipi covers, and many other Indian necessities.*" (Callenbach, 1996, p65).

Culture, therefore, is not only an important part of sustainability, but buffalo are still an important part of life among Native Americans.

Because of the diversity of Tribes, it is difficult to develop standards for the cultural sustainability of Tribal buffalo programs. Although it is difficult, it is the foundation upon which the program should be built. The past wave of buffalo restoration was a grass roots, Tribal member driven effort that wanted to bring buffalo back to Tribal lands and to allow Tribal members access to the animals as they had in the past. There was no effort made to determine which Tribes could receive buffalo. An effort was made to provide animals to all Tribes that wanted buffalo. The grass roots approach included more than just bringing the buffalo back. The restoration of the cultural aspects of the buffalo is very important. Education must be done to teach Tribal members what their buffalo history and culture was, to ensure that the program is being developed to fill the needs of Tribal members and to gain community support for the program. If the program is managed in a way that does not fit the cultural goals of the Tribe, support for the program will suffer and the program may not be sustained.

Educational programs developed to ensure cultural sustainability might include topics from the following list.

1. Tribal Usage of Buffalo
    - Cultural/spiritual purposes
    - Traditional foods including buffalo meat

According to Lulka,

Tribes have developed several distinct distribution channels to disseminate bison meat within their community. Many of these are deeply embedded within the life of the society. In most of these channels, tribes utilize meat from bison raised on their own lands. Bison meat may be distributed during ceremonial events, such as pow-wows or Sundances. Bison may also be consumed during memorials for deceased family members. In some cases, bison meat may be distributed through eldercare facilities and school programs. In each of these examples, the circuit along which bison meat travels is interwoven with the structure of the community. Indeed, the use of bison meat may reaffirm and refortify these social bonds. In addition to these systems, some tribes distribute meat through newly created diabetes programs. These patterns of consumption illustrate the irreducibly social nature of food and diet. And while these tangible networks emphasize the world of consumption, they are frequently matched by a distinct set of tribal production practices because they are still connected with herds in the field. (Lulka, 2006, p.79).

2. Tribal Knowledge of Buffalo
    - Buffalo cultural history and stories
    - History of land tenure
3. Tribal Support of Program
    - Buffalo program activities in the community
    - Buffalo education in schools

Based on past research, education is very important with regards to tribal buffalo programs. The sustainability of tribal buffalo herds and programs isn't based on economics or marketing but on culture. Many tribes possess profound respect for the buffalo as the animal represents more than just a dollar sign; it represents the past, present, and future of many people who once depended on the animal for survival. Education among native and non-native people about buffalo is still a necessity to help people understand the importance and significance of that relationship.

2. **Ecological Sustainability**

Mr. Jim Stone
Dr. Heather Hatfield

The ecological sustainability of the Tribal buffalo programs is an important component of sustainability. Good buffalo management supports cultural sustainability and ensures the integrity of the buffalo as a species. For Tribes that have the buffalo as a foundation of their religion it is imperative the buffalo remains undomesticated, exists in a natural setting, and maintains its genetic integrity.

These principles rely heavily on the "Bison Conservation Management: Proposed Guidelines for Herd Managers" drafted by the World Wildlife Fund's Bison Conservation Working Group.

The three main components of ecological sustainability are:

1. To maintain the wild characteristics of buffalo.
2. Restoration and Management of biodiversity and ecosystem function.
3. Maintaining genetic diversity and integrity.

1. *To maintain the wild characteristics of buffalo*

- Allow buffalo to interact with other buffalo
    - Allowing a natural matrilineal family group structure is very important. Also, maintaining a larger herd allows more social interaction.

- Have mixed age/sex classes
    - Mixed age/sex classes allow younger animals to learn behaviors from older animals. Harvesting of animals based upon reaching a certain age or sex could lead to elimination of some age/sex groups and their behavioral traits. Older animals teach behaviors like wallowing and horn rubbing to the younger buffalo. They also have the social and ecological memory to influence herd social behavior and foraging patterns. Having the mixed age/sex classes allows for competition for food resources, breeding, social hierarchy, and social interactions that favor natural selection instead of human-driven selection.

- Allow buffalo to interact with their habitat
    - Buffalo should be allowed to conduct natural foraging throughout the year. Buffalo should also have the full complement of species that correspond to the natural ecosystem, including predators, parasites and competitors. This allows for natural selection to occur.

- Minimum range size
    - As a guide 5,000 acres is the minimum that would allow buffalo to roam in a natural manner. Smaller pastures can be used in rotation to achieve this number, but

would involve collaborative management for smaller pastures.

- Minimize human handling
    - Limiting round-up to once a year and limiting human contact as much as possible will lead to reduced stress, negative behavioral impacts, injuries and mortality.

2. *Restoration and Management of Biodiversity and Ecosystem Functioning*

- Maintain proven habitat structural diversity
    - To achieve ecological restoration of a buffalo dominated ecosystem, Tribes should attempt to maintain healthy riparian areas, allow for and maintain grass height and species diversity and allow for and (do not intervene) in wallows. These activities will aid in restoration of grasslands and associated species like grassland birds.

- Utilize natural patterns and processes in buffalo range
    - To allow for the establishment of natural processes and patterns Tribes should allow "unmanaged": grazing, fire, native seed dispersal, natural wetlands, wallows and parasites.

- Conserve Species of Interest
    - Tribes should incorporate planning and implementation of species restoration in buffalo pastures.

- Create and implement range biodiversity conservation plans
    - Tribes should attempt to restore native biodiversity.

- Enhance biodiversity on the range
    - Tribes should allow for other species to exist on and have access to range. This may result in competition which would reintroduce some aspects of natural selection.

- Limit herbicide and pesticide usage
    - This should be done in order to limit impacts to non-targeted species and to avoid any negative health impacts.

- Minimize surface water developments
    - This will allow for the restoration of natural hydrology.

3. *Maintaining Genetic Diversity and Integrity*

- Random selection model (harvest or cull)
    - Tribes should limit selection based on specific traits such as aggressive animals and trophy hunting – big bodies,

big heads as this eliminates these traits in successive generations. In the same vein, breeding for specific traits should also be avoided.

- Breeding Competition
    - Breeding competition is important in order to ensure that the best suited animals are breeding and these characteristics are passed on. This means keeping a bull cow ration close to nature - 40% bulls and 60% cows.

- Effective Population Size
    - The effective population size is the total number of breeding individuals, under theoretically ideal conditions, that contributes to the genetic diversity in an isolated population.
    - General recommendations are to assume that the "effective population size" of a conservation herd should be a minimum of approximately 500 breeding animals (this is based on scientific recommendations based on extinction probabilities for isolated populations).
    - Thus, for example, if a bison herd has an adult sex ratio of 1 male: 10 females, a population of about 1,500 adult animals is needed to achieve an effective population size of 540 (a more skewed sex ratio requires a larger herd to achieve a desired effective population size). (Note that our use of the term "herd" should not be confused with

groups of animals that often form within herds and do not consistently associate, but among which breeding generally occurs). For a herd with a ratio of 40:60 male to female, the effective population size would be 100.

- Managing cattle genetic introgression
  - Tribes should attempt to reduce the degree of cattle gene introgression where feasible and where such efforts do not jeopardize bison genetic diversity. Tribes should be cautious in their attempts to reduce levels of cattle nuclear DNA, because animals that are removed from the herd because of cattle nuclear introgression may also possess rare bison alleles that may be lost from the herd. However, removal of animals with cattle mitochondrial DNA haplotypes does not generally pose this risk. Freese et al. states that, "*Some bison populations with low levels of domestic cattle introgression might also be valuable conservation targets due to their historical importance and unique genetic constitution and because they can presumably fulfill the ecological role of bison*" (Freese, et al., 2006, p.179) and "*tribes often take a 'hands-off' approach to bison management and allow, to the extent possible, natural process to unfold. Because many of the tribal bison herds were started or reinforced with surplus bison from national parks and refuges,*

some herds may be free of domestic cattle introgression" (Freese, et al., 2006, p181).

- o Less than 2% cattle genetic introgression can be a guide. This is the federal conservation goal for buffalo herds. Testing of 205 of the herd should be sufficient to determine DNA.

Photo taken by Dr. Heather Hatfield, Modoc Tribe

3. **Economic Sustainability**

Mr. Jim Stone

Most Tribes originally brought buffalo back for the cultural and spiritual importance and they started with limited infrastructure. Usually they had an exterior fence and little else. As Tribes began to utilize buffalo for health purposes and restoring it into the food systems on the Reservations, they had to develop the processes to field kill animals and transport to local processing facilities, and then store the meat while they dispersed it to the Tribal members. Tribes were not anticipating the increased costs associated with the activities. Also, if a Tribe wanted to get buffalo meat into the federally funded program, they would need USDA inspected meat. This meant animals had to be transported live to the USDA facilities, increasing costs for corrals, trucks, trailers, and staff for roundups. These were all costs that the Tribal governments had to cover in order to provide buffalo meat to the Tribal members. This forced some Tribes to develop revenue from the Tribal buffalo herds in order to become sustainable.

The concept of economic sustainability is important. Most Tribes do not generate much revenue outside of land leases, so there is little opportunity for procuring Tribal funds for the buffalo programs complete operation. Having the support of the Tribal members, which can be translated into financial support from the Tribal government, is important for the programs. This can provide a base funding which can then be supplemented with funding from different sources.

The potential for procuring grants exists as a means to support the program as well buffalo marketing activities. Tribes can focus on providing hunts, while keeping in mind the section on mimicking natural selection as best as possible. Tribes can also remove animals for live sale and establish localized markets for sales of buffalo meat for Reservation food systems. The establishment of markets for the sales of buffalo art products and tanned hides and leather is also an option.

ITBC can provide individualized assistance to Tribes looking to develop localized marketing initiatives.

Photo taken by Dr. Heather Hatfield, Spirit Lake

Other Economic Issues

Dr. Heather Hatfield

Funding for tribal buffalo programs can also be a problem. Most tribal programs aren't being funded by the tribe or are only partially funded. Expenses associated with a buffalo program include the following: maintaining fences, equipment, corrals, chutes and pens; paying veterinary fees, the bison manager's salary, and etc. Those costs need to be kept in mind when considering finances.

Fencing

The cost of initial fencing buffalo pasture can be quite high (Table 2). The following information was taken from Government of Saskatchewan website for the cost of a one mile fence composed of barbed wire.

Posts – All posts are six foot and have a spacing of 18 feet. This works out to 293 posts per mile. We assume this length of fence will require 20 brace posts and nine corner and gate posts.

Wire – A roll of barbed wire is ¼ mile long. A three-wire fence one mile long requires 12 rolls and a four-wire fence one mile long requires 16 rolls. Two-strand standard 12.5-gauge barbed wire is used for the barbed wire fences.

Staples – One strand of barbed wire stretched one mile and attached to 293 posts requires approximately 4.3 lbs. of staples. Therefore, a three-wire fence requires 13 lbs. of staples and a four-wire all post fence requires approximately 18 lbs. per mile.

Labor – Erection of fence, including corner posts, bracing, pounding posts, stretching and stapling wire, requires 70 hours per mile for a three-wire barbed fence and 80 hours per mile for a four-wire barbed fence.

Post Pounder – A team should be able to erect one mile of posts in 10 hours.

Table 2: Barbwire fence cost

Materials	**Four standard barbed 2-strand wire all post fence**	**Three standard barbed 2-strand wire all post fence**
Line posts 3-4 inch tops, 6 foot treated	$956.16	$956.16
Brace posts 4-5 inch tops, 8 foot treated	$139.80	$139. 80
Corner and gate posts 5-6 inch tops, 8 foot treated	$86.01	$86.01
Wire		
12.5 gauge, standard 2-strand barbed	$1,012.85	$759.64

Staples, 1.5 inch, 50 pound box	$24.83	$17.94
Tractor, 50 horsepower, total costs	$284.76	$268.94
Post pounder total costs	$176.10	$176.10
Labour	$1,040.00	$910.00
Total Estimated Costs	**$3,720.51**	**$3,314.58**

(Agriculture, 2009, p. 2)

This initial cost does not include the cost of maintenance and is only an estimated cost per mile of fence line. The cost can vary region to region and even city to city for the cost of supplies. It is also recommended that at least a four or five strand wire fence be utilized for buffalo if using barbed wire.

However, please note that no single fence type or design is suitable for all geography, regions, or site conditions. If you plan on having wildlife other than buffalo on your pastures it is also recommended that the bottom and top wires be smooth for wildlife access.

"Some fence designs can disrupt wildlife movements, fragment home ranges, reduce or eliminate access to forage, cover and water, or cause entanglement or impact injuries. . . Pronghorn antelope almost always crawl under fences and rarely jump them. For this species, the bottom strand should be smooth wire set at 16" to 20". Among elk, moose, and deer . . . The bottom wire can be set at 20" high to allow newborn moose and elk to pass, without increasing the risk of bison escaping. Adult moose, elk and deer typically jump fences, but their juveniles tend to go under or through them" (Gates, p.2).

Photo taken by Buffalo Field Campaign, Yellowstone National Park

Corrals

Handling facilities designed especially for buffalo are needed to ensure the safety of both the animals and the handlers. As a rule of thumb, buffalo facilities will work for cattle but cattle facilities won't work for

buffalo. Cattle working chutes, corrals, and fences are generally too small and not high enough or strong enough to handle buffalo. The corral and working chutes need to be at least 7 to 8 feet high.
Catwalks along the outside of chutes and corrals are a good idea for safety. A strong squeeze chute with a crash gate is a must. All gates should be heavy duty, spring-loaded and locking slam gates.

As you are designing your corral system it is important to keep in mind that corners are hard to get buffalo out of and if the alleys are too wide the animals will try to turn and run in the opposite direction. Another important observation is that the less buffalo see the easier they are to work. Make the corral panels a solid piece or cover with dark fabric when working the animals. This has been noticed to have a calming effect on buffalo. The system should move the animals through with a minimal amount of work and stress for everyone. There should also be easy access to all of your pastures and holding pens from the corral system.

One of the best vendors that can be recommended is Pearson Livestock Systems. They make equipment specifically for buffalo and will help you design your corral system with your specific needs in mind. A complete buffalo "model" with tubs, alleys, palpation cage, squeeze box, head-gate, crash-gate, mounts, and buffalo chute costs about $12,000.

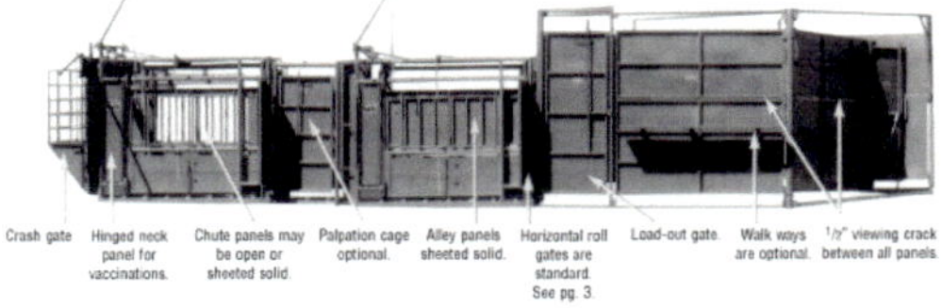

Picture taken from: http://www.pearsonlivestockeq.com/products/bison-handling-equipment/

Tribal Buffalo Meat Sales Program

Mr. Jim Stone

The development of a Tribal Buffalo Meat Sales Program is one that is a very complex endeavor and one ITBC as an organization has struggled with for years. After reviewing the past efforts of ITBC and talking with Tribes and Tribal people about the issue we have developed a model for Tribes to consider when starting a sales program. Time, effort and energy must be spent identifying a workable system because the system is the solution to the problem of how to market buffalo meat. It is also important that the program follows the principles and beliefs of the Tribes and ITBC with regards to our relationship with the buffalo. The core principles are a respect for the buffalo and the utilization of the buffalo by Tribal people. With ITBC this can be seen in our Mission Statement: *"To restore Buffalo to Indian Nations in a manner that is compatible with their spiritual and cultural beliefs and practices."*

For the Tribes the principles and beliefs are unique to each Tribal group and are based on respecting and honoring the buffalo in the management of the herds and in the cultural usages of the animal.

In this model the system will be a utilizing some of the concepts of franchised businesses as the structure of the model and the development of Tribal branding for the buffalo meat products. We would like to utilize the ITBC logo as a national brand while at the same time allowing for the product to be branded with a local Tribe's logo.

The Tribal Buffalo Meat Sales Program is one that is based on the image of a tipi with four poles representing: 1. Product; 2. Service; 3. People; and 4. Finance with the tipi poles being tied together through the Principles and Beliefs of the organizations involved and the system as described in this document.

Background

The economic importance of sustainability for Tribal buffalo programs means making sure that Tribes are able to keep their populations healthy and be able to maintain their herds while generating revenue. Most Tribes started their herds because of cultural purposes and others because they saw a business opportunity. Grass fed buffalo meat is a real commodity in America.

"*Although Indians are not mainly interested in bison for economic reasons, subsistence food production for the Tribe would be certainly welcome. Many Indians have no respect for cattle and believe that eating them makes you weak*". (Callenbach, 1996, p74)

Most Tribal programs aren't being funded by the tribe or are only partially funded. This has been a big issue as it does cost to run a buffalo program. Maintaining fences, equipment, corrals, chutes, pens, vet fees, and even the bison manager salary needs to be kept in mind when considering finances.

Tribes and ITBC have been working on various marketing and sales efforts through the existence of the organization with limited success. The reasons they didn't succeed are the same reason most economic development efforts fail in Indian Country, a lack of proper management, a lack of capital and viable markets for the products/service. The process of selling Tribal buffalo meat products is a complicated activity because of the various entities involved in the project and the lack of developed systems to model a program after.

It was my thought that in order to develop a system that would succeed with the Tribal Buffalo Programs it would have to be designed by Tribal people that have an understanding of how Indian Country operates. I believe it is easier to take a person knowledgeable in Reservation economics and learn the non-Tribal business of buffalo meat sales than to take a marketing expert and teach them how a Tribe functions.

With this idea in mind ITBC has focused our efforts on learning the ins and outs of buffalo meat sales using staff that has a background and understanding of Tribal life. The foundation of this marketing effort focuses on the selling of Tribal buffalo meat into the localized food systems as well as other outlets in order to generate a profit level satisfactory to the Tribe.

The Model

The model has been developed over the last four years as a work in progress with the Flandreau Santee Sioux Tribe in SD and has been modified as needed. Although not a finished product the model will be able to show a format that can be followed by the Tribes. The focus of the program will be to make buffalo meat available in the local food systems at affordable prices. Product will also be avail be for full priced markets. The generation of revenue will then subsidize cultural usages. The image on the following page shows the concept in a visual form.

Photo taken by Dr. Heather Hatfield, Cheyenne Arapaho

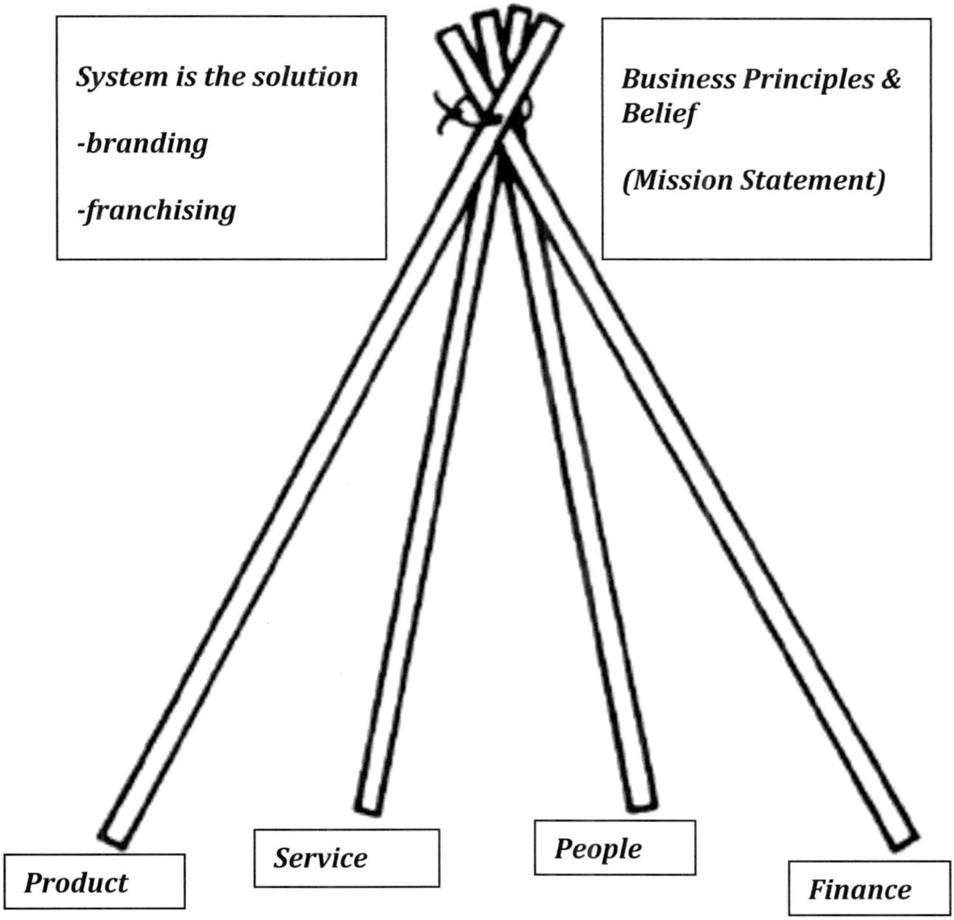
System is the solution
-branding
-franchising
Business Principles & Belief
(Mission Statement)
Product
Service
People
Finance

The image of the tipi was one that came about through conversations with Jerry Noonan about the idea of business management and his usage of a pyramid to details the functions of the various aspects of a successful business. As the tipis poles provide it strength the strength of your business is based on your product, service, people and financial management. And these strengths are held together by the commitment to the system we have developed and the principles and beliefs of the Tribes and ITBC.

Product

The product we are talking about is tribally raised buffalo meat. The product has many positive attributes that can be considered premiums. The product is one of the healthiest meat options available. Grass fed buffalo meat is low in cholesterol, low in fat, high in protein, high in omega 3 fatty acids and low in calories. Buffalo meat has a strong cultural tie with Native Americans and is central to many ceremonies which is why buffalo are raised in a natural system (grass fed, no hormones, etc). This natural production method adds another premium to buffalo meat. In today's society the ability to know the origin of your food and the methods by which it is raised, processed and delivered to the consumer is very important. The idea of Food Security is central to programs like "*Know Your Farmer, Know your Food*", farmers markets and the whole food, slow food movement.

The premiums that are a part of the product are the reason why buffalo meat is so expensive in most markets. The ability of buffalo meat to help to address the current health issues being faced by Tribal people is also an important attribute of buffalo meat. Tribal Members will be getting access to a high quality, high cost and healthy meat product at prices they can afford under the system we are discussing and there will be product available for sale at the higher market prices as well.

Service

When we talk about service in the buffalo meat sales program we are talking about services to the community that we will be providing buffalo meat to. The act of restoring buffalo to Tribal lands and the inclusion of buffalo meat into the diets of Tribal Members is providing a service of cultural restoration. In addition to that we are providing a health service to the Tribal members by giving them access to a locally raised, all natural health meat product. This ability to provide your own food products to your Tribal members also supports the Tribes efforts at creating Food Sovereignty.

People

The people that provide strength to the business are the people who are responsible for caring for the buffalo as well as the people who will be the consumers of the buffalo meat. The goal of the program is to make buffalo meat available to schools, Tribal programs and federal programs at prices they can afford. With a focus on these groups we will be putting buffalo meat into the diets of the youth and the elderly both of which are very important groups in Tribal communities. In addition there will be an increased opportunity for employment as the program grows which is always important in Indian country.

The strength of the program is based in the people committed to providing buffalo meat to the Tribal members and creating a sustainable program that will allow the program to grow. The people that will benefit from the consumption of buffalo meat are important groups within the community which should generate support from programs and individuals to help the program succeed. The human element of the program is a selling point to get people involved and motivated to make the program successful.

Finance

It is important that the program be based on sound financial principles with strong management to ensure the program succeeds. Time needs to be spent researching the potential markets for your product, what prices those markets can pay and what markets the Tribal members and leadership view as most important. The identification of the premiums of the product, the service is provides and the people impacted by the product can be used a selling points to get vendors to buy the product and go out of their way to make sure the product is purchased. As an illustration most Tribes have an issue with selling to their Tribal casinos because for their staff it involves more work than they normally do in procuring products. But if it was explained that the casino purchasing of the product benefits the youth and the elderly by allowing the product to be sold to them at a price they can afford and these may include their kids, parents and other family benefits it is an easier sell. In addition by making the product affordable the consumer benefits by getting a high quality product at a discounted price.

Principles and Belief

The strength of ITBC as an organization is that we have a principle commitment to restoring buffalo to Tribal Lands which is a very strong mission. The Tribes take that commitment and make it their own when they restore the buffalo into their culture and communities. The beliefs both have is that the restoration of buffalo into our daily lives will return our communities back into the strong, healthy people we once were. These principles and beliefs provide a foundation for the manner in which Tribes raise buffalo as wildlife and how the animals is harvested in a respectful and honorable way. It is for this reason that we have focused our efforts on selling buffalo meat in the localized food systems; schools, Tribal programs, local stores and Tribal businesses. Other outlets will be added as needed to ensure financial viability and sustainability of the program.

The principles and beliefs guide the development of the management system and take the strengths of the program; the product, the service, the people and the finances and ties them together with an aspect of Tribalism that makes it unique.

System is the Solution

The goal of the system we have created is to ensure that the strengths of the program are developed in line with and held together by the principles and beliefs of the organizations. The system we have developed takes some of the concepts of branding and franchising into consideration. We would like to utilize ITBC as a nationally recognized brand through our logo and allow the Tribes to develop localized branding by using their logos on products sold in their area. We would like to develop a model for development and management of the program that can be used in the same manner as franchised business. All the necessary development and management formats and steps are delivered to Tribes that want to develop a sales program that includes a menu of the various structures that can be used.

The system will tie the strengths of the program together in conjunction with the principles and beliefs of the participating organizations.

Photo taken by Dr. Heather Hatfield, Nambe Pueblo

Common Diseases

Dr. Trudy Ecoffey

The following diseases are some examples of common diseases that may be found among buffalo populations.

Anthrax

Cause – ingestion of anthrax spores which are found in soil or hay. Can stay in the environment up to 30 years and spores are located in streams, insects birds, dogs and other carnivores. Sometimes frequent rains and washing of soils can expose the bacteria that when ingested by an animal turns into the spores.

Symptoms – usually fatal and causes sudden death; head down and unresponsive; lame, stiff legged and cannot walk; swelling in the underside

Diagnosis – most commonly done by postmortem examination of the carcass. Swaps are taken and sent into a pathology laboratory and should only been done by a survived veterinarian or official from CDC.

DO NOT examine animal by opening and expecting the animal; carcass will bloat quickly after death and has a sawhorse configuration; there will also be signs of dark tarry blood from body orifices; blood of animal may not clot after death.

If anthrax is expected call the local veterinarian immediately or state veterinarian lab.

Treatment – none; may try systemic antibiotics (penicillin's) but outcome is usually death.

Control- This is a reportable disease in United States and Canada. All animals that have been infected will be destroyed and carcasses will be buried. The ground will need to be disinfected and all the animals will be vaccinated and the ranch will be quarantined.

Brucellosis –

Cause – *Brucella abortus*. Can cause disease in cattle, sheep, goats, elk, horses and bison.

Symptoms – In male bison, *B. abortus* infection causes orchitis, seminal vesiculitis and epididmytits. In female bison it causes placentitis and abortion. It may cause abscesses and chronic septic arthritis in both sexes.

Diagnosis – Test done by a blood samples

Treatment- Animals are not treated, but are usually slaughtered.

Control – Routine testing may be done; some states require animals coming in to be tested for brucellosis. Vaccination is not advised because vaccinated animals cannot be distinguished from infected animals.

Tuberculosis

Cause – *Mycobacterium bovis*. Can cause disease in bison and many other animals.

Symptoms – Chronic weight loss, chronic cough and enlarged lymphnodes

Postmortem – tubercular abscesses or granulomas in the lymph nodes of the head and pharynx, and the thoracic and abdominal cavities

Diagnosis – Reportable disease. For bison an intradermal caudal fold skin test is needed.

Treatment – Negative test to the caudal fold skin test after the animals is 18 months old or older

Control – Testing and monitoring of all animals over the age of 18 months. The herd will be then need to be tested on a regular basis to make sure all animals are negative to the caudal fold skin test.

Bovine Viral Disease (BVD)

Cause – BVD is an RNA virus from the family Togaviridae, genus pestivirus and is mostly found in cattle. Bison have been known to have antibody titers to the disease. Transmission of the disease in cattle is direct contact between animals and across the placenta to the fetus. The virus can be shed in nasal discharge, saliva, semen, feces, urine, tears, milk and discharges following abortion of a fetus.

Diagnosis – blood samples for virus isolation or polymerase chain reaction (PCR) testing. Most commonly done postmortem examination and submission of samples to a diagnostic laboratory.

Treatment – no known treatment

Control – Vaccination may provide adequate protection against acute BVD infection, however none have been approved for bison.

Malignant Catarrhal Fever (MCF)

Cause – *Ovine herpes virus* Type 2. The virus that causes malignant catarrhal fever has never been isolated and propagated. The primary host is sheep which MCF causes no problems. MCF can cause disease cattle, bison and deer species.

Symptoms – This disease is known to be fatal in bison. Bison usually die within 7-10 days, and with a chronic form they can die in 156 days, and a recovered form may remain in the bison for an unknown length of time. Clinical signs in bison are reported to include: corneal opacity, conjunctivitis, ocular discharge, nasal discharge, excess salivation, anorexia, diarrhea, melaena, hematuria, multi focal ulceration of the oral mucosa, fever, circling, ataxia, behaviors suggestive of blindness, lameness, and difficulty urinating. The most common signs are anorexia, depression and bloody diarrhea.

Diagnosis – blood samples and nasal swabs that are submitted to a diagnostic pathology laboratory for PCR testing

Treatment – no successful treatment in bison

Control – No commercial vaccinations. Since a large percentage of normal sheep and goats carry MCF it is advisable to avoid any contact between bison and sheep or goats. The safe distance between bison and sheep or goats must be greater than one mile.

Pink Eye –

Cause – Can be caused by *Moraaxella bovis,* but has not been identified in bison.

Symptoms – Increased lacrimation, reddening of the conjunctive, squinting and cloudiness of the cornea, which can led to ulceration of the cornea; temporary or permanent blindness.

Diagnosis – biopsies of the cornea or swabs of the conjunctival sac and sent to a diagnostic pathology laboratory for bacterial culture and identification

Treatment – Small volumes of penicillin and corticosteroids can be injected into the conjunctiva or into the eyelids. Sometimes handling of the bison with pinkeye can be dangerous particularly if the animal is blind in one or both eyes.

Prevention – reduction of face flies. Insecticide ear tags have not been thought to reduce pinkeye in bison. Installing insecticide applicators can be helpful.

Johne's Disease -

Cause – *Mycobacterium aviuam, subspecies paratuberculosis* Most often occurs in white-tailed deer, European red deer, reindeer, sika deer, axis deer, fallow deer, moose, aoudads, mouflons, llamas, African buffalos, water buffalos, camel, stonebucks, antelopes, yaks, gnus, sebu cattle, tule elk, bighorn sheep, mountain goats, American bison and elk as well as cattle, sheep and goats. Very common in cattle and sheep. Most common cause is ingestion of feed or water contaminated by the bacterium.

Symptoms – Has not been reported in bison in recent years. Clinical signs are chronic diarrhea and loss of body weight leading to emaciation.

Diagnosis – Postmortem examination and fecal cultures

Treatment – no known treatment

Control – no known control except depopulate the animals. Advisable to keep bison herds away from captive deer herds and sheep.

Leptospirosis

Cause – many different strains and many different domestic and wild animals can be affected. Usually found in areas with a lot of surface water and source of infection is usually from drinking water that is infected with bacteria.

Symptoms – no known signs in bison, but cattle can show signs of septicemia, interstitial nephritis, abortion and mastitis and high case of fatality rate in cattle.

Diagnosis – Urine can be submitted to a diagnostic pathology laboratory for culture and identification.

Treatment – no known treatment in bison

Prevention – no prevention treatment. There are some vaccines approved for cattle but none for bison.

Mycoplasma Infection

Cause - *Mycoplasma bovis.* Usually occurs in feedlot situations or overcrowding in a small pasture.

Symptoms - Clinical signs included lameness, swollen joints, respiratory distress, and lethargy. Usually older animals (30 months or older) are more susceptible to the infection.

Diagnosis – Taking pathogens for postmortem animals for testing in a pathologic diagnostic laboratory

Treatment- no known treatment

Control – isolate animals that are showing clinical signs of mycoplasma.

Parasitic Diseases of Bison

Anaplasmosis –

Cause – *Anaplasma marginale A. Marginale* parasitizes the red blood cells of the host animal. It is transmitted between host by biting insects. In the US, the most common vector is the tick *Dermacentor andersoni.*

Symptoms – anaemia, jaundice, emaciation and debility. Clinical signs include anaemia, mild depression and listlessness.

Diagnosis – Blood samples submitted to a diagnostic pathology laboratory for a compliment fixation test

Treatment – no known treatment protocols established for bison, however cattle respond to treatment to long acting tetracyclines.

Control – No known control for bison. There are some vaccines available for cattle but not approved for bison.

Disease Associated with Nutrition

Copper Deficiency

Cause – insufficient copper in the diet (primary copper deficiency)

Symptoms – stiff gait, lameness, emaciation, and loss of coat color. Possible low conception rate.

Diagnosis – emaciation, lameness, and diarrheal. Blood test or liver test sent to laboratory

Treatment – dietary supplement of copper; mineral supplement. However, safe levels of copper have not been established for bison. However, 3 to 5% copper in a salt-mineral mix

Control – Feed, water and pasture should be sampled to determine if there is adequate copper available in the diet. Soil samples can be taken in and sent to state or university labs. Also, forage samples can also be tested.

Internal Parasite Control

Dr. Trudy Ecoffey

- Each herd has a unique situation for control of parasites.
- Climate and stocking rate also determine the need for parasite control.
- Dry, cold climates are less likely to have as many parasites issues.
- Low stocking rate, large pastures tend to have less parasites.
- The more confined or the smaller the pasture your bison are on the more likely your bison will have a heavier parasite load. Either way, it is a good idea to control internal parasites that affect your bison.
- If bison are confined (which most are) there is more than likely a need for parasite control.

Grazing density and pasture rotation are very important keys in relation to parasitism. The higher the grazing density, the more likely you will have a high parasite population on the pasture. Animals grazed at low density do not consume herbage close to the ground nor near manure pads, thus ingesting fewer larvae than bison placed on high density grazing. If possible bison should be rotated to new pastures at regular intervals to prevent excessive pasture contamination.

1. Decide if the bison herd needs de-worming. Test fecal samples in the spring or summer when parasite levels are at their highest.

When collecting fecal samples it is best to get a wide sample of pats from different age group of animals. Younger animals tend to have a higher parasite load then older animals.

2. Use rotational Grazing if possible.
3. Minimize stress and use good nutrition.

There are six broad types of internal parasites – stongyles, nematodirus, tapeworms, ascarids, lungworms and coccidiosis.

1. Lungs
    - Lungworm shed larvae on the pasture.
    - Weight loss and coughing.
2. Stomach
    - Ostertagia (eggs resistant to freezing) are one of the few parasites that occur in both adults and young animals.
3. Intestines – nematodirus, copperia
4. Large intestine – coccidian

Clinical signs of parasitism

- Initially no clinical signs
- Poor growth, rough hair coat
- Diarrhea and weight loss (protein loss)
- Part of the herd affected
- Lack of appetite
- Sometimes death

Worms cause the abomasum (the fourth stomach) to not be able to absorb the nutrients Most veterinarians can test fecal samples in their clinics to see what type of parasites are affecting the bison.

Products

Pour-on – (Ivomec[R], Dectomax[R], Eprinex[R])

Injectables (Cydectin[R], Dectomax[R], Ivomec[R])

Feed additive in grain or cake-Safeguard (fenbendazole)

Water treatment- Tramisol (levamisde tablets) of the products for bison have been discontinued, but there are some horse wormers that might be acceptable.

Organic wormers – Diatomaceous earth, herbs, Shaklee Basic H (surfactant)

DO NOT use anthelmintic licks – not safe for bison.

Too often in the past, bison were promoted on the concept that it takes very little feed to keep a bison living. Yes, bison are survivors, but surviving doesn't translate into production. Bison tend to be sensitive to nutritional deficiencies. A weak link in the chain can make the difference between a 50% calf crop and a 90% calf crop. Possibly the most important element of buffalo herd management that can be applied to buffalo is parasite control.

Photo taken by Dr. Heather Hatfield, Pojoaque Pueblo

Stocking Rates

Dr. Heather Hatfield

Dr. Trudy Ecoffey

Stocking rates are expressed as the number of Animal Unit Months (AUM) supplied by one acre of pasture for one year. An animal unit (AU) is defined as one mature 1000 lb cow with or without calf. The animal unit is based on an average daily forage consumption of 26 lbs of dry matter (forage dried until no moisture remains) or "*one AUE consumes about 750 pounds of air dry forage per month*" (Ohlenbusch, 1994, p1). AU/AUM's should be figured out by an experienced range person (in many cases NRCS & BIA will do them for you at no charge).

Let's say that as a buffalo manager you are stocking buffalo in a short grass prairie pasture in eastern Wyoming; a mid-grass prairie pasture in north central South Dakota; and a tall-grass prairie pasture in Iowa. All the pastures are approximately 20,000 hectares. The following will be examples of stocking rate recommendations for each type of pasture, considering that all the pastures are in good to excellent condition, water is available throughout the pastures and the herd that will be stocked as a mix of cows, bulls, yearlings, calves, etc.

The following will be considered in the example: grass type; rotational options; sustainability; weight of animals; marketing strategy and other key elements that might be included in the decision of your numbers.

- 1 bull = 1.3 AUM x 12 months = 15.6 x 750 lbs of forage = 11,700 lbs of forage per year needed
- 1 cow/calf pair = 1 AUM x 12 months = 12 x 750 lbs of forage = 9000 lbs of forage per year needed
- 1 yearling = 0.67 x 12 months = 8.04 x 750 lbs of forage = 6030 lbs of forage per year needed
- Weaned calf = 0.50 x 12 months = 6 x 750 lbs of forage = 4500 lbs of forage per year needed

Suggested forage residue levels for maintaining soil stability and plant vigor		
Vegetation Type	**Pounds/Acre**	**Stubble height (inches)**
Tallgrass	1200-1500	12-14
Midgrass	750-1100	6-8
Shortgrass	300-500	2-3

(Chart taken from R. Lyons & R. Machen publication page 1).

Short grass prairie suggested forage is between 300 – 500 pounds per acre. Considering that the prairie is in good condition we will average 400 lbs per acre. That means an average bull will need about 30 acres in order to survive for a year. Therefore, the stocking rate would be 30 acres per head. You should be able to stock a mixed herd of 1647 animals without harming the land. This is taking the AUM's of the bull into consideration as it is a bigger animal than the cows.

The thought was to take the bigger animal to make the number more conservative on stocking rates so that the land will be able to sustain the buffalo herd year round. This number is not taking the next year's calving rates into consideration. The land should be able to sustain those animals with the thought of rounding up and culling animals once a year. The animals culled can be used for meat purposes for the tribe and/or marketing purposes (by-products or calves).

However, short grass prairie is a type of grassland that is compromised of buffalo and grama grasses. These types of grasses are considered 'increasers' based on the response to grazing pressure. "*Increasers are lower producing, less palatable plants that . . . . tend to increase and take the place of decreasers that weaken or die due to heavy grazing, drought or other range disturbances. The increasers will also be weakened by continued close grazing and decline in abundance*" (Lacey, p2). Taking this into consideration you might to have a more conservative stocking rate. The land should also be cross fenced and animals rotated through pastures to limit the amount of grazing on certain areas.

Mixed grass prairie suggested forage is between 750 – 1100 pounds per acre. Considering that the prairie is in good condition we will average 925 lbs per acre. That means an average bull will need about 13 acres in order to survive for a year. Therefore the stocking rate would be 13 acres per head. You should be able to stock a mixed herd of 3801 animals without harming the land.

A mixed grass prairie has both decreaser and increaser grasses. Decreaser grasses are grasses such as big bluestem or little bluestem and are considered "*high producing, palatable plants that . . . decrease in relative abundance under continued heavy use*" (Lacey, p2). With decreaser plants included in this type of prairie the nutritional value and amount of forage is greater than on a short grass prairie.

Tall grass prairie suggested forage is between 1200 – 1500 pounds per acre. Considering that the prairie is in good condition we will average 1300 lbs per acre. That means an average bull will need about 9 acres to survive for a year. Therefore, the stocking rate would be 9 acres per head. You should be able to stock a mixed herd of about 5491 animals.

The tall grass prairie is mainly compromised of decreaser grasses. The nutritional value is higher on this type of prairie and the forage quantity is greater. However, tall grass prairies are more often overgrazed for those reasons. Again, cross fencing and rotating pastures might be beneficial in keeping the grass from being overgrazed and overtaken by decreasers or invader plants (noxious weeds).

With all three types of prairie types it is important to be "*promoting even livestock distribution by using fencing, salt placement and water development to spread the grazing 'load' over the landscape; avoiding grazing rangeland during vulnerable periods; early spring grazing can stress range plants when energy reserves are depleted as new growth is initiated; and providing effective rest periods after grazing to allow range plants to recover from the stresses of grazing*" (University of Calgary).

Proper stocking is the most important part of successful range management. Conservative stocking can be an effective tool to improve degraded rangelands if they are not heavily dominated by brush.

Also, when assessing pastures take into account that the characteristics of the state obtained from a single source or site are not conclusive for describing that state (of the pasture). In evaluating plant information (grasses), consideration must be given to many factors including:

- Effects of fire or lack of fire
- Impacts of grazing or lack of grazing
- Impacts of rodent concentrations
- Impacts of insects
- Soil erosion or deposition by wind and/or water
- Drought or unusually wet years
- Variations in hydrology and storm events
- Plant disease
- Introduced plant species (including invasive plants and/or noxious weeds)

Keep in mind that these numbers do not consider wildlife or bird populations. In looking at the environment and natural resources as a whole it would be better to have lower numbers to increase nesting birds or other wildlife within the pastures.

The Future of Tribal Buffalo Programs

Mr. Jim Stone

Tribes have been successful at restoring buffalo to Tribal lands. The next step is to integrate buffalo culture into the day to day lives of Tribal communities. At the core of this effort is the restoration of buffalo meat consumption by the Tribal members. Developing a model for getting buffalo from the Reservations and through USDA processing facilities at a cost that is affordable is a major obstacle. One of the biggest battles will be getting USDA to waive or offer a rebate for the inspection costs for Tribal buffalo. The inspection cost is a major component of the costs and places buffalo meat out of the price range of the school system on the Reservations. Another obstacle is getting the Tribes infrastructure developed to produce a consistent supply chain of animals and maintaining the hands off management practices needed to maintain the wildness of buffalo.

The issues that the Tribal buffalo programs face are ones that will force the Tribes to look at the herds that are established as one large herd for management purposes. Animals will need to be moved within the Tribes for genetic issues and cooperatives will have to be made on a regional basis for marketing activities. The Tribes can work cooperatively on some marketing activities like providing buffalo to the US Government for the food Distribution Program on Indian Reservations.

The Tribes and ITBC need to play a stronger role in the management of buffalo under federal jurisdiction. Since the majority of excess federal buffalo are surplused to the Tribes the Tribes should have a say in the management of the resource.

The Tribes and ITBC will need to collectively review the research that has been done on buffalo and determine the suitability of it to the programs the Tribes are trying to develop. The next step would be to identify the gaps in research and to start conducting research that is needed through collaboration with the Tribal colleges.

The Tribes and ITBC will need to develop a program that integrates the work that has been done by individual Tribes and Tribal members with the work that has been done at the Tribal colleges and start from there to build more sustainable buffalo herds. This can be done by their inclusion in ITBC in some manner or fashion.

Photo taken by Dr. Heather Hatfield, Sandia Pueblo

Bibliography

Brown, J. (1953). *The Sacred Pipe: Black Elks' Account of the Seven Rites of the Oglala Sioux.* University of Oklahoma.

Callenbach, E. (1996). *Bring Back the Buffalo: A sustainable fture for America's Great Plains.* University of California Press.

Canby, W. C. (1981). *American Indian Law.* St. Paul, MN: West Publishing Co.

Cunningham, W. P., & Cunningham, M. A. (2008). *Environmental Science: A Global Concern (tenth edition).* McGraw-Hill.

Freese, C. H., Aune, K. E., Boyd, D. P., Derr, J. N., Forrest, S. C., Gates, C. C., et al. (2006). Second Chance for the plains bison. *Biological Conservation* , 175-184.

Gates, C. Cormack. 2006. Fencing guidelines for bison on Alberta public lands with wildlife and access in mind. Faculty of Environmental Design, University of Calgary, Calgary, Alberta, Canada. 8 p.

Halbert, N. D., & Derr, J. N. (2007). A Comprehensive Evaulation of Cattle Introgression into US Federal Bison Herds. *Journal of Heredity* , 1-12.

Lacey, John and John E. Taylor. *Montana Guide to Range Site, Condition, and Initial Stocking Rates.* Montana State University Extension Service. http://msuextension.org/publications/AgandNaturalResources/MT198515AG.pdf

Lulka, D. (2006). Bison and the Food Distribution Program on Indian Reservations. *Great Plains Research* , 73-84.

Lyons, Robert K. and Richard V. Machen. *Stocking Rate: The Key Grazing Management Decision.* Texas Cooperative Extension: The Texas A&M University System. http://agfacts.tamu.edu/D11/NLO/Pasture%20Management/StockingRatePasture.pdf.

Ohlenbusch, Paul and Steve L. Watson. *Stocking Rate and Grazing Management.* Kansas State University Agricultural Experiment Station and Cooperative Extension Service. February 1994. http://www.ksre.ksu.edu/library/crpsl2/MF1118.pdf

The Editors of Time-Life Books. (1993). *The American Indians: The Buffalo Hunters.* Alexandria, Virginia: Time Life Books.

University of Calgary. *Fencing Guidelines for Bison on Alberta Public Lands.* http://www.canadianbison.ca/producer/Resources/documents/BisonGuidelinesJune30_2006_Web.pdf

About the Writers

Dr. Heather Hatfield is currently the Technical Services Provider for InterTribal Buffalo Council and a tribal member from Bay Mills Indian Community. Heather received her B.S. degree in Environmental Science from Oglala Lakota College in 2003; M.S. in Biological Sciences from South Dakota State University in 2006; and Ph.D. in Biological Sciences from South Dakota State University in 2012. Her dissertation focused on the sustainability of tribal buffalo programs. She can be contacted at heathermhatfield@hotmail.com or heather@itbcbison.com.

Dr. Trudy Ecoffey has a background in Range Ecology, Bison Management and Wildlife. Dr. Ecoffey has worked for the Oglala Sioux Parks and Recreation Authority for the last 8 years and previously worked at the Oglala Lakota College in the Ag and Natural Resources Department, South Dakota Extension Program and ITBC. Dr. Ecoffey obtained her PHD from South Dakota State in Biology with a dissertation on bison and range. Her master's degree is from the University of Nebraska – Lincoln in Range Management with an emphasis in wildlife where she wrote a bison management curriculum, and her BS Degree is also from UNL in Animal Science. Dr. Ecoffey has been working on swift fox translocation project, range surveys, bighorn sheep and endangered species on the Pine Ridge Reservation in southwest South Dakota. She currently is working on writing a comprehensive wildlife management plan for the Oglala Sioux Tribe. She resides near Wounded Knee, SD with her three children and enjoys working with horses, hiking and spending time with her family. She can be contacted at trudymillerecoffey@outlook.com.

Mr. Jim Stone is currently the Executive Director of InterTribal Buffalo Council (ITBC) and a tribal member of Yankton Sioux Tribe. Previous to working at ITBC, Jim worked for the Yankton Sioux Tribe where he served as the Director of the Environmental Program, Tribal Utilities Program and was the Chairman of the TERO Commission. Jim received a B.S. degree in Biology and a minor in Chemistry from the University of South Dakota in 1993. He can be contacted at jstone@itbcbison.com.

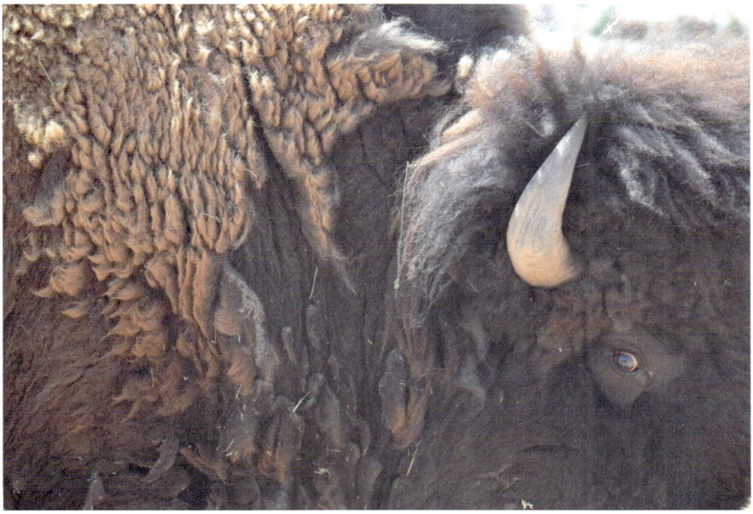

Photo taken by Dr. Heather Hatfield, Sandia Pueblo

ITBC information

InterTribal Buffalo Council (ITBC) has a membership of 58 tribes in 19 states with a collective herd of over 15,000 buffalo. Membership of ITBC remains open and there is continued interest by non-member tribes in the organization. ITBC is committed to re-establishing buffalo herds on Indian lands in a manner that promotes cultural enhancement, spiritual revitalization, ecological restoration, and economic development.

ITBC has been reorganized as a federally chartered Indian Organization under Section 17 of the Indian Reorganization Act. This was approved by the Bureau of Indian Affairs in 2010. ITBC consists of a membership that includes all the tribes that have joined the organization and is governed by a Board of Directors, which is comprised of five elected officers and four regional representatives.

The role of ITBC, as established by its membership, is to act has a facilitator in coordinating education and training programs, developing marketing strategies, coordinating the transfer of surplus buffalo from national parks and tribal lands, and providing technical assistance to its membership in developing sound management plans that will help each tribal herd become a successful and self-sufficient operation.

Made in the USA
Monee, IL
05 November 2026

67378942R00043